Creating a Pagan Theocracy: Modern Science

The following essay will be Chapter Sixteen in the forthcoming book *The Political Theory of Christ: And Its Creation of Our World.*

Today's histories of modern science begin with Galileo and Newton. This is not so much an error as it is a complete misunderstanding. It represents the attempt, which is common enough in writing history, to read something into the past that is simply not there. And what is not there, in the seventeenth and eighteenth centuries, is *modern science*.

What really began with Galileo and Newton – and with Descartes – was faith in *reason*. This faith in reason, of course, would eventually lead to a faith in modern science, but the two beliefs should not be conflated. For the early modern belief in reason and the later modern belief in science are not at all the same beliefs.

As late as 1800, most Western intellectuals considered reason to be a form of divine revelation. They believed, for example, that reason revealed the true nature of God. For as late as 1800, most Western intellectuals still believed in God and in the existence of a reality beyond the senses, and believed that this reality was revealed by reason. By that date, of course, many Western intellectuals were no longer orthodox Christians. But few were atheists and none were agnostics, a word that would not be invented until 1869. Although, by the year 1800, most Western intellectuals considered "experimental knowledge" to be the most potent form of reason, they also believed that such knowledge described only the material aspects of reality. Reason, on the other hand, comprehended reality as a whole.

If one were to pick the year in which modern science was born that year might be 1833, since that is the year when the word *scientist* was coined. In short, before 1833, there were no scientists in the world, which is just another way of saying that there was no *science* in the world, at least as we use that word today. And even in 1833, the word scientist was viewed by its inventor as something of a nonsense word. It was as if, today, someone invented the word "knowledgist" to describe someone who specializes in "knowledge as such." Since it is obvious to us that there can be no such thing as *specializing* in "knowledge as such," the word is clearly nonsense. And this is why calling someone a scientist in 1833 was also clearly nonsense. For the word science, at the time, still had only one meaning: "systematic knowledge." Every form of systematic knowledge was a science. Philosophy was a science. Theology was a science.

But if there were no scientists before 1833, there were people who were called "natural philosophers." And a natural philosopher is exactly what those words imply, someone who philosophizes about nature. And this is why today's histories of science, which so confidently refer to modern science when speaking of Galileo or Newton, are anachronistic. Before 1833, there were no scientists, which means that there was no modern science.

However, by 1890, or just sixty years later, a majority of Western intellectuals had come to believe that "experimental knowledge" (now called "modern science") was the only real knowledge. All other forms of knowledge were now regarded as being merely opinion or superstition. Experimental knowledge (i.e. modern science) had displaced reason as the full and final revelation of reality. And this new revelation had nothing to do with discovering the true nature of God or understanding the nature of the reality beyond the senses. To the contrary: modern science was a completely material revelation of a completely material world. By 1890, for many Western intellectuals, faith in modern science was now inseparable from the belief that there was no reality except material reality.

Even those intellectuals whose academic disciplines were non-scientific now struggled to discover some method by which they could turn their disciplines into a form of experimental knowledge. Philosophers, who had once believed that philosophy was the queen of sciences, because philosophy was said to define reality, now completely gave up this belief and attempted to turn philosophy into a form of experimental science. Liberal theologians engaged in a similar reconstruction of Christian theology. They argued that Divine revelation was to be found, not primarily in the scriptures, whose narratives were "pre-scientific," but in the human *experience* of the divine. And this experience could only be properly understood *scientifically*.

In the first half of the twentieth century, the American ruling class overwhelmingly continued to believe in God, although they increasingly believed in the god of liberal theology. The largest and most influential Protestant denominations came under the control of religious liberals. Still: the great mass of the American people remained orthodox in belief, including most of those who belonged to the liberal denominations. And since, during this period, the American political order was still largely local and democratic, American law and social mores continued to reflect a local, democratic, and *Christian*, consensus on the necessary relationship between morality and law. The Christian secular republic of the American nineteenth century was definitely on the wane. However, during the first half of the twentieth century most Americans, including most religious liberals, still considered America to be a Christian society.

But the influence of liberal theology on the American elite now meant that questions of morality and law would be increasingly framed in terms of a *scientific* understanding of morality, rather than according to a *Christian* understanding. But, for the moment, America remained a visibly *Christian* secular society.

All of this came to an end in the second half of the twentieth century. The rise of the power of the American national state after 1932 effectively ended local democracy as the primary law-making authority in America. Control over the law now passed to the national state and to the least democratic branches of that state: the judiciary and the bureaucracy. The Christian secularism that had once defined American life, from the adoption of the Constitution until well into the nineteen fifties, was then formally outlawed by the Supreme Court during the nineteen sixties. Following the cultural revolution of that same decade, effective control of the American national state began to rapidly pass to a secularizing – *to a paganizing* – elite.

These two events alone, the judicial and cultural revolutions of the nineteen sixties, would ensure that the American political and social order would now cease to be Christian. But the word *pagan* could never be used to describe this revolution, since the overwhelming majority of the American people continued to identify as Christians. Thus the word *secularism* became the necessary euphemism that would be used to describe this transformation. It was now said that America was a "secular society" and that this had been the original intent of America's Constitution. The reality was that America was in the process of ceasing to be a *Christian* secular society and was in the process of becoming a *pagan* secular society. It was only that the word pagan could never be used to describe this process.

But the decade of the sixties was witness to more than just the final overthrow of America's Christian secular republic. It was also witness to the beginnings of an explicitly pagan intellectual revolution that would overthrow the authority of modern science.

Modern Science: Decline and Fall

In the nineteen fifties, philosopher of science Karl Popper, in a widely discussed book, argued that the rules of science were more like the rules of chess than the rules of logic. Popper held that the rules of science were simply the agreed-upon rules about how science was to be conducted and that these rules did not inhere in the nature of reality itself. Popper also argued that while a scientific theory could always be proven to be false, no scientific theory could ever be finally proven to be true. This was because there was always the possibility that additional evidence might falsify even the most well-established theory. While Popper agreed that science progressively revealed reality, he argued that what was being revealed was not reality itself, but only "increasing information" about reality. Popper believed that the ultimate nature of reality might well be beyond the ability of science to discover at all.[1]

---

[1] Karl Popper, *The Logic of Scientific Discovery* (1959)

Also in the fifties, philosopher of science Michael Polanyi argued that scientific knowledge, like all other forms of knowledge, was largely *tacit* in nature. By this, Polanyi meant that every scientific theory implicitly contained non-scientific assumptions – *tacit assumptions* – that could never be scientifically proven. Therefore no scientific theory was a completely objective understanding of reality. Polanyi also argued that scientific theories were the products of particular individuals working within particular traditions of science. Although science aimed at an objective understanding of reality, neither the individual scientists, nor the scientific traditions in which they worked, were wholly objective. Thus, Polanyi argued, scientific theories should primarily be understood as forms of "personal knowledge," the personal knowledge of the individual scientist who was working within a particular scientific tradition.[2]

However, Popper and Polanyi were just the prelude to the coming revolution. In 1962, Thomas S. Kuhn published a small book, *The Structure of Scientific Revolutions*, which precipitated an intellectual civil war which continues to this day. Kuhn's basic argument was that scientific theories changed for *non-scientific* reasons. His analysis of the history of scientific revolutions claimed to demonstrate that scientists created new theories, and overthrew old ones, based upon a personal adherence to some particular *paradigm* of science. And such paradigms, Kuhn argued, were never believed in for completely rational reasons. Instead, they were believed in, at least in part, for non-rational reasons.[3]

Kuhn's intent in writing had not been revolutionary. But he had crossed a line that Popper and Polanyi had refused to cross. Although they had both argued that scientific theories

[2] Michael Polanyi, *Personal Knowledge: Towards a Post-Critical Philosophy* (1958)
[3] Thomas S. Kuhn, *The Structure of Scientific Revolutions* (1962)

were influenced by non-scientific factors, they continued to view science as an objective attempt to understand an objective reality. But Kuhn had depicted scientific theories as being decisively influenced by non-rational beliefs, with each succeeding scientific paradigm being rationally *incommensurate* with every preceding paradigm. Because Kuhn had not intended to deny scientific objectivity, he was at first taken aback by the vociferous objections to his book by those who called it "anti-scientific." He was taken further aback by those who hailed his book as a manifesto that revealed the starkly subjective and social character of modern science.[4]

To understand the full impact of Kuhn's book today, it is necessary to understand what most Western intellectuals believed modern science to be in 1962. Beginning in the late nineteenth century, the dominant philosophy of science was called *positivism* (which then became *logical positivism*, which then became *logical empiricism*). Positivism held, more or less, that there was a one-to-one correspondence between established scientific theories and reality.

However, within ten years of the publication of Kuhn's book, there were almost no positivists left in the world. The paradigm had changed. Even those who continued to uphold the superiority of modern science to other forms of knowledge now generally conceded that a scientific description of "reality as such" was probably beyond the methods of science. They also conceded that the creation of scientific theories was not an entirely objective enterprise. But most physical scientists, as well as most philosophers of science, continued to hold the belief that scientific theories corresponded, more or less, to something called reality. This new and chastened view of science came to be called *scientific realism*. However, scientific

---

[4] John H. Zammito, *A Nice Derangement of Epistemes: Post-Positivism in the Study of Science from Quine to Latour* (2004), Chapter 3.

realism was the least radical response to the revolution unleashed by Kuhn's book.

The most radical response was found among those who would later be called *postmodernists*. The postmodernists argued that modern science was primarily a "discursive construct." By this, they meant that modern science was primarily a manipulation of language in service to the belief system of the dominant political and social class. To the postmodern intellectual, the primary purpose of science was not to explain the cosmos, but to control social meaning on behalf of the ruling class. Because most postmodern intellectuals also believed that there was no such thing as reality, but that there were only "constructions" of reality, modern science was viewed as just one more "construction." Worse, modern science was considered to be the dominant modern construction of reality and thus a primary source of political and social oppression.

Although most postmodern intellectuals believed that the physical cosmos did in some sense exist, they also believed that the cosmos, as described by modern scientists, was largely an artifact of hegemonic rule.

By the late twentieth century, postmodernism had become dominant in the universities, at least outside the physical sciences. A majority of Western pagan intellectuals now believed that postmodernism described the only reality there was. In the view of the postmodernists, although modern science had gotten rid of God, which was all to the good, this had been accomplished only by turning science into a form of divine revelation. The claim of modern science to be the full and final revelation of reality simply substituted a false scientific revelation for the equally false Christian revelation. However, there were no revelations of reality, since there was no reality. There were just constructions of reality.

Now the claim of modern science to explain reality rested primarily on a belief in something called *reductionism*. Reductionism held that every event that occurred in the cosmos could ultimately be *reduced* to the primary physical forces of the cosmos. Modern science thus rested upon the belief that everything that occurred in nature had no intrinsic reality of its own, but was just an accidental by-product of these primary physical forces. Human life, for example, originated out of non-life.

Postmoderns believed in reductionism, but of a radically different kind. The postmodern intellectual held that all the individual realities in the cosmos could be ultimately reduced, not to primal forces of physics, but to the collective interaction of those individual realities.

Indeed, one prominent postmodern theory of science took the name "Actor in the Network Theory" or ANT, for short. According to ANT, there were actors – or, more precisely, there were *actants*, since an actant could be a machine, a physical process, or an animal, as well as a human being – and then there was the network. The network was simply the evolving, dynamic relationships of all the actants. It was these two things together – the "actants in the network" – that constituted the full description of reality. In place of the scientific belief that matter/energy determined everything in existence, the postmoderns believed that it was the actants in the network who "constructed" the only reality there was. The physical cosmos was only the background noise to this more primal reality. According to the postmoderns, to understand ANT – or something very much like it – was to understand the only reality that could be known.[5]

Bruno Latour, one of the creators of ANT, is probably the best known postmodern philosopher of science. It is central to

---

[5] Zammito, 183-202.

Latour's thought that he makes no distinction between *force* and *reason*. Latour argues that anything that can be said about the use of force may be equally said about the use of reason. Modern science believed that reason represents the force of *truth*. But since there is no such thing as truth, this can only mean that *reason* represents *force*. Latour also argues that no distinction can be made between science and technology. He therefore speaks only of *technoscience* since, in his view, the authority of science is indistinguishable from the technology that science creates to enforce that authority. It is the purpose of *technoscience* to create a technology to "explain" – *to control* – reality. Shorn of its technology, science has no political or social authority. This is why there is no such thing as science, but only techno-science. In Latour's noted formulation: "Science is not politics. It is politics by other means."

An Expanding Post-Scientific Intellectual Revolution

However, the struggle over the meaning of modern science has turned out to be far more complex than just a showdown between scientific realism and postmodernism. Indeed, these two philosophies are best understood as opposing poles in a much wider conflict. Between these two poles lies a whole range of philosophies of science. Although most of these philosophies point to one pole or the other, many are not so easily classified.

Take philosopher of science Bas van Fraassen. Van Fraassen argues that science must be understood as a completely objective methodological enterprise and that it cannot be understood in any other way. This would seem to point him in the direction of scientific realism. But van Fraassen also argues that science is not a description of reality, which seems to point him in the direction of postmodernism. When properly understood, however, he belongs in neither camp.[6]

Van Fraassen argues that modern science constructs objective models of reality, but that these are models of what is *observed* about reality, rather than models of reality itself. Van Fraassen contends that the idea of reality is a meta-physical, rather than a scientific, concept. Thus science, in his view, does not address questions of reality at all. In the words of Bas van Fraassen: "[T]he aim of science is not truth as such but only *empirical adequacy*, that is, truth with respect to the observable phenomena."[7] It is the "observable phenomena," as modeled by objective theories, which define science. Therefore, by definition, no scientific theory ever corresponds to reality. Science is the construction of *models* of reality.

Van Fraassen also argues that there are no scientific "laws of nature." Modern science, he argues, by claiming to have discovered the laws of nature, merely took over from Christianity a belief in the "laws of God" and inappropriately applied that metaphysical belief to its physical models. But since modern science only creates models of *observed* phenomena, these models can only reveal *regularities* in what is being observed. They cannot model something that is called a law. All that a model can establish is that certain causes, in our experience, always produce certain effects. There is no "law" establishing that those causes must always produce those effects.[8]

Another philosopher of science, Ronald N. Giere, whose work overlaps van Fraassen's, argues that it has become necessary for today's philosophers of science to create a *via media* between scientific realism and postmodernism (which Giere labels "social constructivism"). Giere calls this proposed

---

[6] Bas Van Fraassen, *Scientific Representation* (2008).
[7] Bas Van Fraassen, *Laws and Symmetry* (1989), 192.
[8] Van Fraassen (1989), Parts I and II.

via media "scientific perspectivism."[9] He argues that it is the perspective of the individual philosopher of science that alone determines the relative weight that the individual philosopher will assign to any claim of scientific objectivity versus an opposing claim of social construction. And this individual perspective is inevitably the product of both the subjective and objective beliefs of the individual philosopher. And these beliefs can never be fully differentiated as being entirely subjective or objective.

Scientific perspectivism, Giere argues, is not a theory about how science operates. It is the practical recognition that science is a matter of personal judgment. So long as an individual philosopher's perspective concerning the meaning of a particular theory can be rationally defended, his perspective should be treated as legitimate. This remains true even if fundamental disagreements over the truth of a particular scientific theory arise from a clash of differing philosophical perspectives.

Ronald N. Giere:

> If scientific knowledge is perspectival, scientific claims are neither as objective as objective realists think nor as socially determined as even moderate constructivists often claim….perspectival realism is as much realism as science can provide. Objective realism cannot be even an ideal goal.[10]

But Giere is postmodern enough to assert that scientific perspectivism includes the repudiation of all final truths, whether the final truths of scientific realism or the final truths of Christianity. Indeed, Giere considers scientific realism to be

---

[9] Ronald N. Giere, *Scientific Perspectivism* (2006).
[10] Giere, 16.

little more than a religious faith, left over from the nineteenth century belief that modern science was the full and final revelation of reality.

Beyond Reduction: The Increasing Disunity of Science

One of the few agreements among philosophers of modern science today is that there is no such thing as *reductionism* in science, at least to any significant degree.[11]

Reductionism is the belief that all the theories found in the various fields of science can ultimately be "reduced" to the theories found in more "fundamental" fields of science. This usually entails the belief that all theories in science can ultimately be reduced to the theories of physics. Take, for example, scientific theories of human consciousness. First, human consciousness is defined by the theories found in the science of psychology. Second, those psychological theories are then said to be "reducible" to the theories found in biology and chemistry. Third, the theories found in biology and chemistry are said to be reducible to the theories found in physics. Thus human consciousness is ultimately nothing more than a special case of physics, meaning that it is ultimately nothing more than the product of matter/energy. And this means that there really is no such thing as human consciousness, except as a kind of illusion.

Under this conception of science, which was dominant from the late nineteenth to the middle of the twentieth century, the unity of science was the product of the reducibility of all theories ultimately to the theories of physics. Today, however, most philosophers and historians of science no longer believe that this kind of inter-theoretic reductionism

---

[11] For an overview see Steven Horst, *Beyond Reduction: Philosophy of Mind and Post-Reductionist Philosophy of Science* (2007), Chapter 3.

exists. And the reason why they have ceased to believe in reductionism is because, since the nineteen sixties, both analytical and historical studies of how science actually operates have revealed huge explanatory gaps between the theories found in the various sciences. Few real inter-theoretic reductions can be shown actually to exist. They are just assumed. Thus the conclusion of most of today's philosophers and historians of science is that inter-theoretic reductionism is nothing more a presupposition of a now discredited understanding of science, rather than something that can be scientifically demonstrated to exist.

Philosopher of science Steven Horst:

Enlightenment rationalists and Logical Positivists favored a reductive metatheory, but did so largely on armchair, aprioristic grounds. To the extent that one has reason to trust armchair reasoning to lay down the norms for the shape of the sciences, one *might* even still be inclined to view this as a tenable *normative* project. But to the extent that the philosophy of science is guided by a careful examination of how real science is done, this metatheoretical picture does not stand up to much scrutiny.[12]

But if there is no reduction of the theories of the various sciences to the theories of physics, then there is no unity of science. And if there is no unity of science, *then there is no science*. There are merely the individual sciences, with each being bound by the rules and traditions peculiar to that particular science.

This, at least, was the conclusion that most philosophers and historians of science had reached by the beginning of the

---

[12] Horst, 60-61.

twenty-first century. It was only practicing scientists, together with a minority of philosophers of science, who continued to argue for the unity, and thus for the existence, of something called *science*.

In the space of a single century, from 1890 to 1990, Western intellectuals had come full circle. In 1890, most Western intellectuals had been eager to discover some method by which they could turn their non-scientific academic disciplines into a form of experimental science, since they believed experimental science to be the only real knowledge. By 1990, however, most Western intellectuals, outside the hard sciences, were now eager to demonstrate that their academic disciplines constituted knowledge that was *independent* of the claims of experimental science. The 19th century Western pagan belief that modern science was the full and final revelation of reality – thereby displacing the Christian revelation – was, by the end of the twentieth century, now just a minority belief among Western pagan intellectuals.

The Religious Character of All Theories

At this point, we could conclude this chapter. We have traced the decline and fall of modern science as a source of authority among Western pagan intellectuals. The consensus among philosophers of science, by the beginning of twenty-first century, was that there was no such thing as science, but that there were only the sciences – with each science being governed by its own rules and traditions. Moreover there was a general consensus that the various sciences did not describe reality as such, but that they were only models of reality. Modern science had failed as the full and final revelation of reality. It had failed as the intended revelation that was to have succeeded the Christian revelation.

However, we will not be ending the chapter at this point. Throughout this book we have emphasized that there is only one reality, which means – among other things – that all politics is religious and that all religion is political. We will now expand this thesis to include the observation that all the sciences are religious, by which we mean that all scientific theories, at bottom, are based upon some particular religious belief.

This is not a view that is widely held today, but that cannot be helped. We make this assertion, not because it is widely held, but because it is true.

And we turn to philosopher of science Roy A. Clouser as our guide in this matter, since he is the philosopher who has most closely examined the religious character of all theories, including all scientific theories.

Roy A. Clouser:

> [This is not] the claim that the proposals of theories are all deduced from religious convictions (though that has happened at times). Rather, I mean that some religious belief or other delimits an acceptable range of interpretations of the nature of whatever a hypothesis proposes. It is in this sense that I find the influence of religious belief to be utterly pervasive.
>
> And it is in this sense that virtually all the disagreements between rival theories in the sciences and in philosophy can ultimately be traced back to the differences between the religious beliefs that guide them. [13]

---

[13] Roy A. Clouser, *The Myth of Religious Neutrality: An Essay on the Hidden Role of Religious Belief in Theories* (2005), 3.

Clouser argues that every theory rests on what he calls a "divinity belief." This is not necessarily a belief in God. Rather, it is a belief in the existence of a "non-dependent ultimate reality" that determines the nature of the rest of reality. It is the functional equivalent of a belief in God, which is why Clouser labels it a divinity belief.[14]

Clouser argues that there are three dominant divinity beliefs in our time: theism, paganism, and pantheism.

Theism is defined by Clouser as the belief that a Creator and a creation constitute reality. For the theist – for Jews, Christians, and Muslims – the "non-dependent ultimate reality" that creates and sustains the rest of reality is God. The cosmos is God's creation and is completely dependent for its continuing existence upon God.

Paganism is defined by Clouser as the belief that there is only one reality, the cosmos, and that some aspect of the cosmos determines the nature of the rest of the cosmos. For example, the ancient pagans believed in many gods, but their underlying belief was in the existence of a cosmic order of which the gods were a part. The gods did not determine the cosmic order, but were determined by it.

Modern Western pagans who are scientific materialists hold a similar belief. Like the ancient pagans, the materialist believes that the cosmos is all there is. He also believes that a "non-dependent ultimate reality" within the cosmos (usually, matter/energy) is that aspect of the cosmos that determines the nature of the rest of the cosmos.

Pantheism is defined by Clouser as the belief that there is a single, undifferentiated reality, which is then masked by the multitudinous *seeming* realities that make up the world. These many seeming realities are just illusions that hide this single reality. The ultimate truth is that "All is One." Human beings, animals, plants, the galaxies, are merely surface illusions of

[14] Clouser, Chapters 1-6.

this underlying unity. Pantheism can take more than one form. In Hinduism, it can be a belief in an actual deity that constitutes the single reality. In Buddhism it can take the form of atheism or the belief that a radical "absence" or "void" is the single, undifferentiated reality.

Now for the reader to fully engage Clouser's argument, he should turn to Clouser himself. Here we are interested solely in the examples that he provides to show that most modern scientific theories rest upon pagan divinity beliefs. In short, it is not just that modern science points away from Christian belief. Rather, it is that modern science is largely a *pagan* system of belief. To demonstrate this, Clouser provides a close analysis of three scientific disciplines: mathematics, physics, and psychology. He shows not only that the theories found in these disciplines are constructed according to various pagan divinity beliefs, but that these three sciences are riven by competing pagan divinity beliefs.

Pagan Divinity Beliefs in Mathematics

Today, most people believe that mathematics has nothing to do with religion, since "one plus one must equal two" no matter what one's religious beliefs. But as Clouser points out, once we get beyond the basics of "abstracting and symbolizing quantities and noticing the most obvious laws that hold among them," the question then becomes: "what, exactly, do the symbols of the formula represent? In other words, what is a number? And as soon as this issue is raised, we find that there are serious disagreements among the mathematicians."[15] These disagreements are inherently religious.

One ancient mathematical divinity belief is called the Number World Theory. This was the belief held by the

---

[15] Clouser, 131

Pythagoreans and by Plato, and is also held by some modern mathematicians, which rests on the divinity belief that "the world of mathematical entities is not only real, but more *real* than the things that we observe as objects existing in space and time." The physical cosmos is believed to be nothing other than a physical instantiation of mathematical principles. Thus mathematics determines the nature of reality.[16]

Against this divinity belief, the nineteenth philosopher John Stuart Mill argued that it is only our *sensations* of reality that are real. Since we can know nothing about reality except what our sensations tell us, our sensations are the only reality we can know. Numbers, Mill argued, have no intrinsic reality, but are simply generalizations about our sensations. Mills argues that the formula "one plus one equals two" is true only because our sensations tell us that it is true. But it is possible that, at some point in our experience, that "one plus one" might *not* equal two, because our sensations might tell us that it equals something else. Our uniform experience, of course, is that one plus one is always two. However, this is only our experience. Since we cannot know the totality of all our possible experiences, our experience – at some point – might include the datum that one plus one does not equal two. In other words, Mill argued, mathematics is not divine and therefore does not determine reality. Mathematics is simply a method by which we organize our sensations, which alone constitute reality. This was John Stuart Mill's pagan divinity belief.[17]

Now against the Pythagoreans and John Stuart Mill, the philosopher Bertrand Russell argued that it is *logic* that determines reality. Russell rejected the Pythagorean belief that mathematics was divine by arguing that mathematics was just a shortcut to logic. In Russell's view, "all of math is either

---

[16] Clouser, 133-134
[17] Clouser, 134-135

identical with, or derivative from, logic." Russell then argued, against Mill, that logic also determines the nature of our sensations. One plus one will always equal two, no matter what our sensations may tell us, because logic – and not our sensations – determines reality. It is logic that is divine. This was Bertrand Russell's pagan divinity belief.[18]

Now against the Pythagoreans, Mill, and Russell, the philosopher John Dewey argued that mathematical symbols stand for precisely *nothing*. According to Dewey's divinity belief, mathematics is neither true nor false, but is *useful*. There is no such thing as "true" or "false." Dewey's belief was that human beings are completely biological entities attempting to survive in a completely physical environment, and that this is the full description of reality. Because human beings, alone among the animals, have intelligence, they are capable of altering their environment through the creation of tools. Mathematics is a tool that human beings have created to manipulate their environment. And this means that numbers possess no reality beyond their use as tools. According to John Dewey, "to say that something is true means no more than that it works." This was John Dewey's pagan divinity belief.[19]

There are other mathematical pagan divinity beliefs. Mathematical intuitionists, for example, similar to the Pythagoreans, believe that mathematics determines reality. But they decisively break with the Pythagoreans since they subordinate logic to mathematics, rather than mathematics to logic. According to this divinity belief, "if logical paradoxes arise concerning a mathematical system, that is a problem for logic" and it is not a problem for mathematics. And because mathematics is not bound by logic, reality is not bound by logic. Based upon this premise, mathematical intuitionists

---

[18] Clouser, 135-136
[19] Clouser, 136-138

reject an entire branch of modern mathematical theory, "the theory of transfinite numbers developed by George Cantor." This theory, which is regarded by most modern mathematicians as the most important mathematical advance of the past century, is regarded by mathematical intuitionists as complete nonsense.[20]

Historian of mathematics Morris Kline:

> The current predicament of mathematics is that there is not one but many mathematics and that for numerous reasons each fails to satisfy the members of the opposing schools. It is now apparent that the concept of a universally accepted, infallible body of reasoning – the majestic mathematics of 1800 and the pride of man – is a grand illusion...The disagreements about the foundations of the "most certain" science are both surprising and, to put it mildly, disconcerting. The current state of mathematics is a mockery of the hitherto deep-rooted and widely reputed truth and logical perfection of mathematics.[21]

Pagan Divinity Beliefs in Physics

Most people today believe that modern physics has nothing to do with religious belief, since the same physical laws must apply to everyone. But Clouser demonstrates, through an examination of three of the giants of modern physics – Ernst Mach, Albert Einstein, and Werner Heisenberg – that each of these men held radically different pagan divinity beliefs. And

---

[20] Clouser, 139-142
[21] Morris Kline, *Mathematics: The Loss of Certainty* (1980), 6, quoted in Clouser, 141.

these radically different beliefs caused them to have radically different views of physics.[22]

Ernst Mach, like the philosopher John Stuart Mill, believed that only sensations were real. Mach, together with a large body of other scientists and philosophers at the beginning of the twentieth century, considered atoms and sub-atomic particles to be nothing more than "useful fictions." They were just a way to talk about reality, without being real, because they could not (at that point) be perceived by scientific instruments. Any description of reality that treated atoms and sub-atomic particles as if they were real entities, existing independently of our sensations, was treated by Mach as nothing more than an unsubstantiated hypothesis.

Albert Einstein, while appropriating Mach's insights, decisively rejected his pagan divinity belief. Einstein believed that there was a reality external to the senses, although he agreed with Mach that this could not be scientifically demonstrated. But Einstein's justification for believing in the existence of an external reality was that such a belief decisively aided the mind in understanding the nature of reality. And this signaled to Einstein that an external reality probably existed. He also believed that this external reality was wholly governed by mathematics and logic. This was Einstein's pagan divinity belief. Although Einstein sometimes publicly spoke as if he believed in God, the word "God" for Einstein was merely a metaphor for the underlying Reason that he believed to be reality.

Now against Mach and Einstein, Werner Heisenberg held that elementary particles were, in a sense, physically real, but that they were also partly mathematical entities not governed by logic. According to Heisenberg – and this was basis for his understanding of quantum mechanics – although it is possible

---

[22] Clouser, 147-159.

to calculate both the position of a particle and its momentum, neither can be calculated simultaneously. Heisenberg stated that this was not because we lacked the technical ability to do so. Instead, Heisenberg argued, elementary particles simply do not *have* a position and a momentum simultaneously. It is only our act of measurement that *causes* the particle to possess either a momentum or a location. In Werner Heisenberg's own words: "this is a very strange result since it seems to indicate that [our] observation plays a decisive role in the event and that the reality varies, depending on whether we observe it or not."[23]

Logically speaking, this is nonsense. As a matter of logic, a particle must have both a position and a momentum nor can this reality be changed by our observation. But the logical contradiction disappears once we accept Heisenberg's divinity belief that elementary particles are partly mathematical entities that are not bound by logic. And this is why Einstein could never reconcile himself to the dominant interpretation of quantum mechanics, since his divinity belief was that reality was physical and logical. Einstein's pagan divinity belief was fundamentally at odds with Heisenberg's pagan divinity belief.

Roy A. Clouser:

> Perhaps it is now clear that even though all these thinkers claimed to accept atomic theory, they mean something very different by it – so different that it is fair to say that the twentieth century has actually produced three atomic theories, not minor differences within one and the same theory. For Mach, atomic theory meant inventing a system of micro-entities that

---

[23] Werner Heisenberg, *Physics and Philosophy* (1958), 52, quoted in Clouser, 155.

is useful though populated with fictions. For Einstein, it meant postulating purely physical objects which we never experience. For Heisenberg, it meant postulating micro-entities that comprise reality and that, while composed of physical energy, are essentially mathematical in nature. These sharp disagreements …rest on different views of what is divine.[24]

Pagan Divinity Beliefs in Psychology

Psychology is Roy A. Clouser's third example of how pagan divinity beliefs operate in modern science. Clouser examines behaviorism, which was the dominant psychological theory in America during the first two thirds of the twentieth century.

In the late nineteenth century, the new science of psychology was initially believed to consist of the study of the "mind" or of "consciousness." This belief was then completely superseded, early in the twentieth century by a radically reductive view. The new view held that concepts like "mind" or "consciousness" were much too subjective to be scientific. Psychology was therefore redefined as the science of *behavior*, since behavior could be objectively measured. According to this belief, the science of psychology *was* the study of measurable behavior.

Eventually, two schools of behaviorism contended with each other. The first school focused on the study of individual behavior, while the second school focused on social behavior. These were not overlapping approaches, but were mutually exclusive approaches. They were mutually exclusive because each school rested upon a radically different pagan divinity belief.

---

[24] Clouser, 157.

Those who believed that psychology was the study of individual behavior held that all human behavior was ultimately reducible to biology and physics. This prompted the question of whether psychology could be called an independent science at all, rather than as being merely a sub-discipline of biology. But as Clouser notes, the response of the behaviorists to this fundamental question was to ignore it. It was simply asserted that psychology was an independent science, even though it was completely reducible to the sciences of biology and physics.[25]

One possible way out of this theoretical difficulty might have been for the behaviorists to believe, with John Stuart Mill and Ernst Mach, that reality could be defined as our *sensations*. This belief would have allowed them to construct a science of psychology that was independent of biology and physics, since their psychological theories could be reduced to theories of *sensation*. However, as a matter of history, the behaviorists were convinced that their science was entirely reducible to biology and physics. And this is why they concluded that there was no such thing as the human "mind" or "consciousness." The human experience of having a mind, they argued, was a subjective illusion masking a entirely biological process.

J.B. Watson, the founder of behaviorism, argued that "all subjective terms such as sensation, perception, image, desire, purpose, and even thinking and emotion" needed to be purged from the vocabulary of the scientist.[26] His replace-ment for these subjective terms was found in the single phrase: "stimulus and response." Stimulus and response explained the whole of human behavior. Watson considered behavior to be nothing more than a complex series of reflex actions. Behavior was simply the automatic response to

---

25 Clouser, 163.
26 J.B. Watson, quoted in Clouser, 164-165.

biological programming. Human beings had no free will or individual consciousness, except as an illusion. According to Watson's pagan divinity belief, human behavior *was* "stimulus and response."

Other psychologists engaged in variations on this theme. E. M. Thorndike argued that individual behavior also took place in response to what he called "reinforced" and "aversive" stimuli. But these concepts were rejected by Watson as being too subjective and unscientific.[27] B. F. Skinner then developed a theory based upon what he called "operant" behavioral response. Skinner believed that it was only by a complete understanding of the full history of the conditioned responses of any individual that a genuinely scientific description of that individual's behavior was possible.[28]

Roy Clouser:

> Common to all these theories is the total rejection of allowing into psychology anything about human mental life that is prima facie non-behavioral, such as thoughts, feelings, purposes, and even perceptions.

> Even this brief summary should be sufficient to establish that something very odd is going on. Since all of us constantly experience our own thoughts, feelings, perceptions, intentions, etc., why are these ignored by psychology? …Why [do] behaviorists regard thoughts and perceptions as [unwarranted] *assumptions*?[29]

---

[27] Clouser, 165-166
[28] Clouser, 166.
[29] Clouser, 166.

The behaviorist view, Clouser argues, ultimately derives from a rejection of the concept of free will. Indeed, free will must be rejected by the behaviorist, since if human beings are really free to behave as they choose there can be no *science* of human behavior. According to the behaviorists, psychology is a science only because it describes a completely closed system that can be predicted (at least in theory) at every point. Thus human behavior *must* be completely determined by the brute facts of biology and physics – *and by nothing else*. Behavioral psychology describes a completely physical process devoid of purpose or intent.

Now one might ask why any human being would want to believe that his behavior is completely determined. One might even want to ask how one can possibly know this to be true, since our belief that it is true would also be completely determined. Weren't the behaviorists the victims of their own illusion?

Roy Clouser:

> …[W]hat makes the theory of behaviorism attractive to its advocates is not its explanatory power, since it is patently incoherent. Rather, its attractiveness stems from a particular vision of what science *should* be, which is based in turn on a specific view of the nature of reality and divinity…. The real explanation of [this view] is that it is rooted in the religious belief in the divinity of matter/energy. This is the driving motive of the perspective, and the real source of its power over those who do science under its direction.[30]

---

[30] Clouser, 170.

By eliminating everything that any normal human being would view as part of being human – free will, perception, feelings, and purpose – behavioral psychology invokes a pagan divinity belief in an entirely natural process, in which human beings are completely the products of the interaction of energy and matter. According to this divinity belief, no scientific explanation of human behavior can involve human purpose or free will, since this would mean that human beings are more than simply the random products of matter and energy – and of evolution. And since, *by behavioral definition*, human beings cannot be more than this, there can be no such thing as human free will or purpose.

This brings us to the second school of behavioral psychology, which rejected this divinity belief. According to the second school, although biology and physics set decisive limits on human behavior, all behavior is – at bottom – a form of *social construction*. Thus human behavior, instead of being completely reducible to the principles of biology and physics, is also reducible to those sociological laws that have been discovered to govern societies. Social behaviorism decisively broke with a purely physical understanding of behavior. It assumed that society was, at least in part, an entity that was *independent* of biology and physics.

Alfred Adler and Eric Fromm are the two psychologists who exemplify this pagan divinity belief. Both men were neo-Marxists.[31] This is not surprising, since Marxism also rests upon the belief that humanity is a social construction, although Karl Marx explained this construction entirely in terms of economic relationships. The social behaviorism of Adler and Fromm was an updating – but also a repudiation – of Marx's pagan divinity belief. Both Adler and Fromm argued that it was society, rather than economics, that created human behavior.

---

[31] Clouser, 171-180

Adler broke with the understanding that human behavior was completely reducible to biology and physics by asserting that psychology was a social science possessing its own standards of evidence, which were independent of the physical sciences. Although Adler agreed that behavior was largely determined by biology and physics, he argued that everything specific to being human was socially determined. In Adler's view, all individuals naturally strive for superiority over all other individuals, but are limited in their striving by the existence of those other individuals. Society therefore exists to define the limits of permissible behavior. It is "the logic of communal life" that defines what it is to be human.

Adler held that Marx was the first theorist to actually understand that humanity was a social construct. But Adler turned Marxism on its head. Beginning his intellectual life by holding the Marxist belief that history was completely determined by economic relations, he eventually became convinced that economic relations were themselves determined by the historical development of societies. Adler thus made social determinism, rather than economic determinism, the foundation of human behavior.

Eric Fromm was more explicitly Marxist. Fromm continued to believe that economics completely determined the nature of society, but also came to believe that society determined the nature of the family, which then determined the nature of the individual. For Fromm, psychology was the scientific study of a chain of social determinism, beginning with an analysis of economics at the top of the chain down to an analysis of individual behavior at the bottom. Fromm believed that although economics, society, and the individual could be analyzed as separate categories ultimately they were one category. Fromm also imported into his social theory the explicitly Marxist belief that history was leading to the creation of a global socialist utopia.

However, by the end of his life, Eric Fromm decisively rejected this understanding. He came to believe that human life possessed an inner dynamism of its own, a dynamism that was somehow independent of biology, physics, and even society. The human claim to know truth, as well as the human ability to act upon that claim, was somehow a reality that existed on its own. Human beings, Fromm now believed, were free. However, at the same time, he also continued to believe that all human behavior was completely determined. Logically speaking, this was nonsense. One was either free or determined; logically one could not be both simultaneously. Fromm admitted the logical contradiction, but insisted that it was the belief in "logical contradictions" that was false. He maintained that it was possible to believe both in human freedom and in a completely determined human behavior. To him, these were merely different ways of talking about the same thing.

Fromm now turned to the religions of the East in support of this belief. Like many Hindu thinkers, and almost all Buddhist and Taoist thinkers, Fromm came to the belief that human existence was a matter of "both is and it is not." The East had always recognized that logical contradictions were just an illusion. Having begun his life as a Marxist, Fromm ended it as a pantheist.

Robert Rosen versus Modern Biology

In any battle between two opposing armies there is always a center of gravity in each of the opposing forces, one of which must be broken if one side or the other is going to win. Biology is the center of gravity for the modern pagan understanding of science. In none of the other physical sciences are pagan assumptions about the nature of reality so deeply rooted. The theory of evolution plays the central role in this, but only because that theory is the product of fundamental pagan assumptions concerning the nature of reality.

Today there are alternative voices to the reigning beliefs of modern biological science, with the most prominent being the Intelligent Design movement.[32] However, our critique of modern biology will be guided by the theoretical biologist and biophysicist Robert Rosen, who had no interest in questions of design and who was committed to a completely pagan understanding of biology. Nevertheless Rosen, who died in 1998, rejected the central pagan divinity belief of most of today's biologists. Rosen's background was in mathematics and systems theory, and he brought that background to bear on what he considered to be the fundamental theoretical error of modern biology.

Probably Robert Rosen's most important book was *Life Itself: A Comprehensive Inquiry into the Nature, Origin, and Fabrication of Life*. In this study, Rosen begins by noting that modern biology rests upon a reductionism that is anchored in physics. It is therefore physics that provides all the scientific principles from which inferences are made by biologists about the nature of life. Modern biologists simply assume that all organisms can be completely understood in terms of the physical parts that make them up. Thus life is understood

---

[32] See, for example, Stephen C. Meyer, *Darwin's Doubt: The Explosive Origin of Animal Life and the Case for Intelligent Design* (2014)

according to the metaphor of the machine. The principles of physics, although formulated to describe inorganic matter, are also assumed to completely describe organisms.

Robert Rosen:

> According to this view, there *is* no other science than physics; everything that we call a science is ultimately a special case of physics.[33]

This, then, is the "divinity belief" of most modern biologists. Indeed, most of today's biologists are atheists and materialists, since that is the nature of their religious commitment. Rosen rejects this conception of biology. He begins by asking what the physicists themselves have said about the relationship of physics to biology.

Robert Rosen:

> Living things are surely material; they are manifestations of matter; surely then the secrets of matter must contain the secrets of life. Surely, the physicist, who is concerned with matter in all of its manifestations, will have eagerly striven to translate insights about matter in general into corresponding insight into matter's greatest mystery.[34]

Except that it turns out that the physicists haven't done this. Physicists are almost completely uninterested in "matter's greatest mystery." And Rosen knows why: theoretical physics

---

[33] Robert Rosen, *Life Itself: A Comprehensive Inquiry into the Nature, Origin, and Fabrication of Life* (1991), 3-4.
[34] Rosen, 11-12.

is focused on "the universal and the general." And in the view of most physicists, biological organisms are nothing more than "special systems" that are *somehow* the product of these universal and general laws. But physics as such has nothing to say about biology. Since most of the cosmos is not alive, since it is inorganic, the existence of life is simply irrelevant to the problems of physics.

Rosen notes that a few physicists, Walter Elsasser for example, regard this attitude as more than a little peculiar. Elsasser argues that, although organisms are extremely rare within a largely inorganic cosmos, this does not necessarily mean that they must be described by the same general laws as inorganic matter.

Elsasser points out that anything rare in physics disappears into the averages. Thus life disappears into the averages. But although, he argues, organisms cannot violate the general laws of physics, perhaps it is the case that organisms are governed by other laws not derivable from these general laws. This understanding, of course, contradicts the divinity belief held by most modern biologists that organisms are *nothing more* than specialized examples of the laws of physics. But what if they really *are* more than this?[35]

Robert Rosen:

> On the face of it, there is no reason at all why "rare" should imply anything at all; it needs to be nothing more than an expression of how we are sampling things, connoting nothing at all about the things themselves… Why could it not be that the "universals" of physics are only so on a small and special (if inordinately prominent) class of material systems, a class to which

---

[35] Rosen, 12-13.

organisms are too *general* to belong? What if
physics is the particular, and biology the
general, instead of the other way round?[36]

And that would mean, Rosen recognized, a revolution in
biological theory. And such a revolution would only
moderately impact physics. Modern physics, Rosen points
out, has in the recent past already undergone similar
revolutions, being forced to accommodate the "phenomena of
electricity and magnetism" and of "spectra and chemical
bonding." The real revolution would occur within biology.
Among other things, the Darwinian claim that physical
evolution completely determines biological change would be
overthrown. For if organisms could be described by laws that
were peculiar to themselves, then evolution – assuming that it
even exists – would become largely an external history of an
internal process.

Robert Rosen:

[Modern] biologists believe that life is
somehow the *inevitable* necessary consequence of
underlying physical (inanimate) processes; this
is one of the well-springs of reductionism. But,
on the other hand, modern biologists are also,
most fervently, evolutionists; they believe
wholeheartedly that everything about organisms
is shaped by essentially *historical, accidental*
factors, which are inherently unpredictable and
to which no universal principles can apply. That
is, they believe that everything important about
life is *not* necessary but contingent....What is

---

[36] Rosen, 13.

relinquished…is any shred of logical necessity in biology, and with it, any capacity to understand. [37]

As a theoretical biologist, Rosen's main purpose was to lay the groundwork for a revolutionary new kind of biology. In Rosen's view, biology needed to be re-founded upon the irreducible complexity – *not the design* – inherent in the organization of living things. This complexity, Rosen argued, could only be described according to rules governing the complexity itself in its "relational effects." No biological entity, Rosen argued, is entirely explicable in terms of the physical parts that make it up, which is why life must be analyzed according to the relational effects that constitute the actual organization of living things. Only by formulating rules governing those organizational effects could a new science of biology be created.

Now our purpose here is not to endorse Robert Rosen's radical assault upon the theoretical foundations of modern biology. As the reader might suspect, his influence is quite limited among modern biologists, although he does have adherents. Our purpose, rather, is to show that the pagan divinity belief of most modern biologists – that life is entirely explained as a specialized example of the laws of physics – is simply that: *a belief.* Robert Rosen provides one way forward to create a new science of biology, based on the very different – although in his case still pagan – belief that life can be explained according to independent laws that describe its complexity. Such laws, according to Rosen's pagan divinity belief, remain entirely physical and do not imply design. But the information that they contain transcends any purely evolutionary explanation of reality. Indeed, even to posit the existence of such laws is to herald the end of Darwinian evolution as the primary explanation of life.

---

[37] Rosen 13-14.

Summary and Conclusion

Every society that has ever existed, as well as every society that will ever exist, will always be founded upon some religious belief. Christianity, even now, is the self-declared religion of most Americans, although from the middle of the nineteenth century it has been pagan theories of political and social life that have dominated American society.

But even as Christianity ceased to be the religious basis of that society, the pagan believers in modern science found themselves confronted with their own spiritual crisis. The nineteenth century belief that modern science was the full and final revelation of reality was, by the late twentieth century, now only a minority belief among pagan intellectuals, at least outside the physical sciences. Beginning in the nineteen sixties, an intellectual revolution erupted that would convince most philosophers and historians of science that there was no such thing as *science*, but that there were only the *sciences*. And the sciences no longer explained reality itself, but only modeled that reality and, further, modeled only those aspects of reality that could be measured according to the current practices and traditions of the particular sciences.

By the end of the twentieth century, modern science had ceased to be the religion of most Western pagan intellectuals. But if modern science was not to be the new religious basis of society, another pagan faith would be needed. And this new pagan faith, based upon the belief that there was no such thing as reality, but that there were just "constructions" of reality, had a name: *postmodernism*.